STUDENT GUIDE

MAKING MATHEMATICAL ARGUMENTS

# GENERALIZING ABOUT NUMBERS



MathScape™
SEEING AND THINKING MATHEMATICALLY

What is involved in writing a mathematical argument?

# MAKING MATHEMATICAL ARGUMENTS.

## PHASE**ONE**
Signs, Statements, and Counterexamples

When a statement about numbers is always true, we call it a rule. One way to show that a statement is true is to model it with objects. In this phase, you will use cubes to help you think about statements about adding, subtracting, multiplying, and dividing with signed numbers and answer the question: Is it always true?

## PHASE**TWO**
Roots, Rules, and Arguments

Another way to show that a statement is always true is to look for special cases. In this phase, you will examine different mathematical arguments about squares, cubes, and roots to look for special cases (0, 1, proper fractions, and negative numbers). This prepares you to write your own mathematical arguments.

## PHASE**THREE**
Primes, Patterns, and Generalizations

Patterns can be used as a basis for a rule and to explain how you know a rule is true beyond just giving examples. By the end of this phase, you will choose an interesting pattern (or possible pattern) and experiment with it. These patterns involve squares, cubes, primes, factors, and multiples.

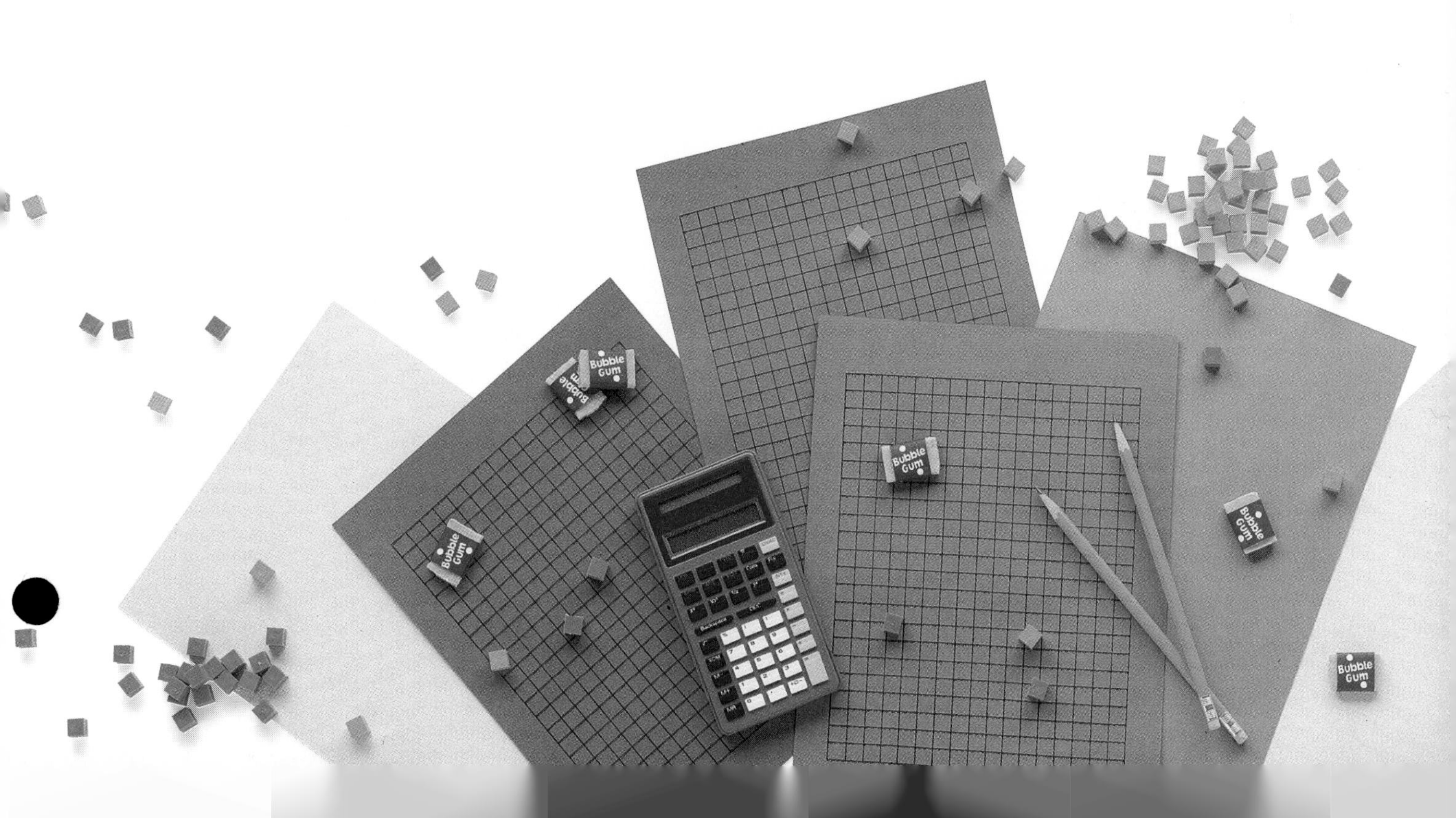

# PHASE **ONE**

A counterexample shows that a statement is not always true. For example, a counterexample to the statement "All apples are red" is the statement "Some apples are green." Why might knowing how to come up with a counterexample be useful?

Understanding and using signed numbers is an important skill. In what professions might you use this skill?

# Signs, Statements, and Counterexamples

## WHAT'S THE MATH?

*Investigations in this section focus on:*

### NUMBER and OPERATIONS

- Understanding signed numbers
- Understanding how to add, subtract, multiply, and divide integers

### MATHEMATICAL REASONING

- Making and evaluating mathematical statements about positive and negative number operations
- Looking for counterexamples to determine whether a mathematical statement is untrue

### NUMBER SYSTEMS

- Understanding how addition, subtraction, multiplication, and division are related to one another

mathscape2.com/self_check_quiz

# 1 Statements About Signs

EXPLORING INTEGER STATEMENTS WITH COUNTEREXAMPLES

**You know the rules for adding and subtracting whole numbers so well that you hardly have to stop and think about them.** In this lesson, you will use examples and counterexamples to explore statements about adding and subtracting with signed numbers.

## Use Cubes to Model Calculations

How could you use cubes to model adding and subtracting positive and negative numbers?

Using cubes can help you get a better sense of how to calculate with positive and negative numbers. After the class discusses the handout Cube Calculations with Signed Numbers, complete the following:

1. On your paper, write each of the Signed Number Problems shown. Use cubes to figure out the solution. Then write the solution.
2. Make up a problem in which you need to add zero-pairs to your cubes to solve it. Start with 5 cubes. Illustrate your problem with cubes.
3. Make up a problem in which you need to add some negative cubes to solve it. Illustrate your problem with cubes.

**Signed Number Problems**

**1.** $6 + (-3)$ **2.** $-5 - (-4)$ **3.** $5 - (-3)$ **4.** $-4 + (-2)$
**5.** $-6 - 3$ **6.** $-3 - (-5)$ **7.** $5 + (-7)$ **8.** $-4 - 2$
**9.** $2 - (-1)$ **10.** $-4 + 7$

## Look for Counterexamples

How can you argue that a mathematical statement is always true, or show that it is not always true?

For each statement below, decide if the statement will always be true. If the statement is not always true, show an example for which it is false (a *counterexample*). If the statement is always true, present an argument to convince others that no counterexamples can exist.

1. I tried four different problems in which I added a negative number and a positive number, and each time the answer was negative. So a positive plus a negative is always negative. — Hyun

2. I noticed that a negative number minus a positive number will always be negative, because the subtraction makes the answer even more negative. — Tanya

3. I think that a negative number minus another negative number will be negative. because with all those minus signs, it must get really negative. — Hyun

4. A negative decimal number + a positive decimal number will equal 0 because they will cancel out. One example of this is $-0.25 + 0.25$. — Tanya

5. A positive fraction, like $3/4$, minus a negative fraction, like $-1/2$ will always give you an answer that is more than 1. — Hyun

6. A negative decimal + a negative decimal will always give you a negative answer. — Tanya

7. You never need to add zero-pairs to your cubes when doing an addition problem — Hyun

signed numbers
counterexample

page 124

# 2 Counterexamples and Cube Combinations

PREDICTING RESULTS OF INTEGER ADDITION AND SUBTRACTION

**In the last lesson, you explored statements about the results of adding and subtracting with signed numbers.** In this lesson, you will analyze equations you create yourself. This will help you make predictions about whether the result of an equation will be positive or negative.

## Use Cubes to Create Equations

How can you use cubes to create signed number problems when you know what the answer is?

The value represented by a given number of cubes depends on how many of the cubes are positive and how many are negative. In the following problems, build representations with cubes, and then record each equation in writing.

1. Make all possible combinations of 3 cubes. For each combination, bring in additional cubes so that the overall total is $-4$.

2. Make all possible combinations of 4 cubes. For each combination, remove cubes or bring in additional cubes so that the overall total is $-1$.

3. Make all possible combinations of 5 cubes. For each combination, remove cubes or bring in additional cubes so that the overall total is $-1$.

### Signed Number Problem

Start with 6 negative cubes. Show a problem with the cubes for which the answer is $-3$.

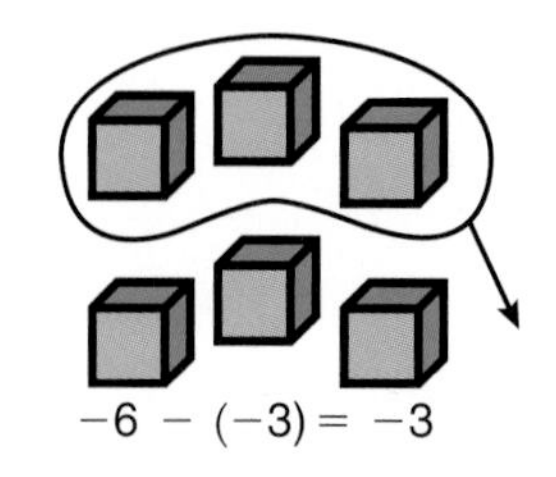

## Sort the Solutions

Investigate the questions below about the results of adding and subtracting signed numbers. As you investigate, keep in mind that important question: Is it always true?

How can you predict whether an answer will be positive or negative?

1. What are the possible combinations of positive and negative numbers in an addition or subtraction equation that involves just two numbers? One example is positive + positive = positive. Be ready to share all the combinations you can think of in a class chart.

2. Look at the class chart of combinations.

   a. For which combinations on the chart can the sign of the result always be predicted?

   b. For which combinations on the chart does the sign of the result depend on the particular numbers in the equation?

Remember to look for counterexamples if you think you have found a rule.

## Explain the Results

Choose one kind of combination from the class chart that will always have a negative answer. Choose another kind of equation from the class chart of combinations where the sign of the answer depends on the numbers being added or subtracted.

For both combinations, write one paragraph that explains the following:

- Will the result always be positive or negative? Why?
- If the result can be either positive or negative, what does it depend on?

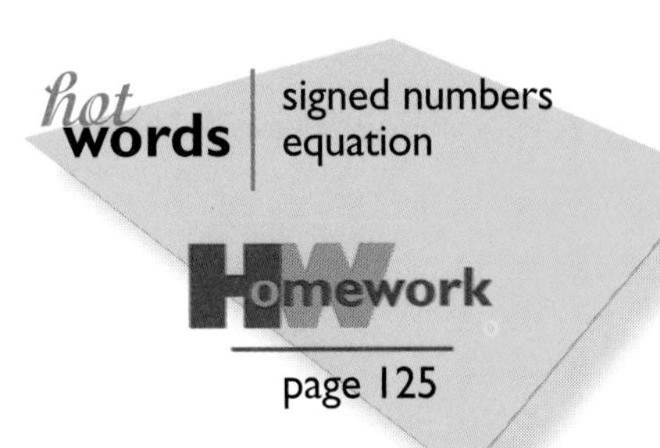

# 3 More Cases to Consider

EXPLORING INTEGER MULTIPLICATION AND DIVISION

**Now that you have learned how to add and subtract signed numbers, it is time to move on to multiplication and division.** After finding the rules for multiplication, you will investigate rules for division. Then you will be ready to write your own statements and counterexamples to summarize what you have learned so far about signed numbers.

## Use Cubes to Model Multiplication

**How can you use cubes to model multiplication of signed numbers?**

You have used cubes to model adding and subtracting with two numbers and made a chart of the results. Can you find ways to use cubes to model multiplying different combinations of signed numbers?

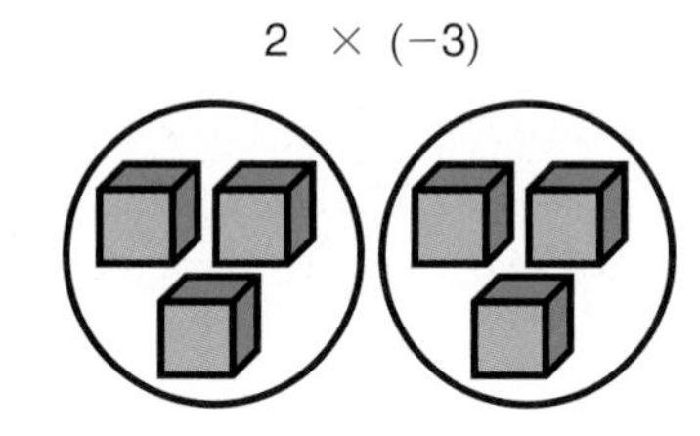


1. List all the different combinations of two signed numbers that could be multiplied.
2. For each combination you listed, write down some sample problems.
3. Try to find a way to use cubes to model examples of a multiplication problem for each combination of two numbers. Draw pictures of your models to record your work.
4. Write down any conclusions you can make about the results when multiplying each combination.

## Investigate Division of Signed Numbers

How can you use what you know about multiplication to think about division of signed numbers?

Now that you have used cubes to model multiplying different combinations of signed numbers, the next step is to think about division. You may use cubes and what you know about the relationship between multiplication and division as you work on these division problems.

$$-6 \div (-3) = ?$$

1. List all the different combinations of two signed numbers that could be used in a division problem.
2. For each combination you listed, write some sample problems. Think about what you know about the relationship between multiplication and division before you write down the answers to your sample division problems.
3. Write down any conclusions you can make about the results when dividing each combination.

## Create Statements and Counterexamples

In your group, make up eight statements about operations with positive and negative numbers. Four of your statements should always be true. Four of your statements should not always be true and should have counterexamples. Try to include some statements that are true in some cases and others that are always false. Make sure you include all four operations in the eight statements: addition, subtraction, multiplication, and division.

On a separate sheet of paper, make an answer key for your statements that shows:

- the statement
- whether it is always true or not always true
- one counterexample, if the statement is not always true

# 4 Rules to Operate By

SUMMARIZING RULES FOR OPERATING WITH INTEGERS

**In this lesson, you will think about rules for adding, subtracting, multiplying, and dividing with signed numbers.** As you think about what operations might be equivalent and about counterexamples, you will be summarizing what you have learned about signed number operations in this phase.

## Find Operations that Are Equivalent

**Which operations with signed numbers are equivalent?**

**1** Using 3, −3, 5, and −5, write as many different addition and subtraction problems as you can that have the answer 2 or −2.

a. Look at the problems you wrote and think about when adding and subtracting are equivalent, or when you get the same result.

b. Write statements about when you think adding and subtracting are equivalent. Will the statements you have written always be true? Can you find any counterexamples?

**2** Write down all the multiplication and division equations you can using all combinations of two numbers from 3, −3, 5, −5, 15, and −15. Use the equations you write to help you complete the following:

a. Write a general rule for multiplication and division problems that have positive answers.

b. Write a general rule for multiplication and division problems that have negative answers.

## Find Counterexamples

How can you apply what you know about integer addition and subtraction and counterexamples?

Look at the statements in the box below.

- If you can find a counterexample, write the statement and its counterexample.

| Is It Always True? |
|---|
| Positive + Positive = Positive |
| Positive + Negative = Negative |
| Negative + Negative = Negative |
| Positive − Positive = Positive |
| Positive − Negative = Positive |
| Negative − Positive = Negative |
| Negative − Negative = Negative |

## Write and Test Statements About Multiplication and Division

**1** For each problem below, tell whether the answer will be positive or negative and how many negative numbers are multiplied.

a. $2 \times (-2)$

b. $2 \times (-2) \times (-2)$

c. $2 \times (-2) \times (-2) \times (-2)$

d. $2 \times (-2) \times (-2) \times (-2) \times (-2)$

**2** Write a statement that tells whether your answer will be positive or negative when you multiply several numbers together.

**3** Can you find a counterexample to the statement you wrote about multiplication? If so, rewrite your statement so that it is always true.

**4** Using what you have learned about multiplication, write a statement that tells whether your answer will be positive or negative when the problem uses division.

page 127

# PHASE TWO

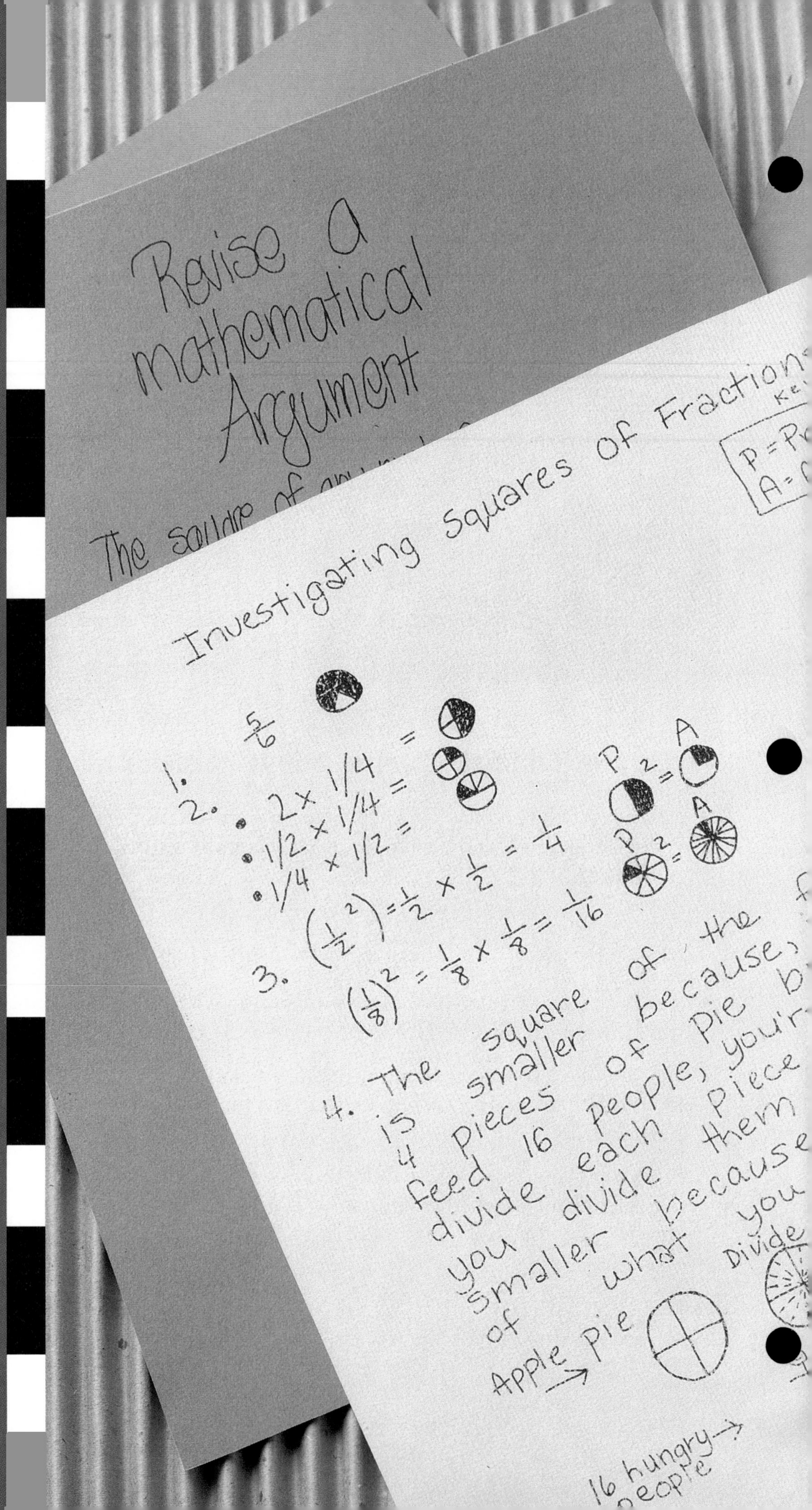

Revise a mathematical Argument
Investigating Squares of Fractions
1. 5/6
2. • 2 × 1/4 =
• 1/2 × 1/4 =
• 1/4 × 1/2 =
3. (1/2 ^2) = 1/2 × 1/2 = 1/4
(1/8)^2 = 1/8 × 1/8 = 1/16
4. The square of the
is smaller because,
4 pieces of pie
feed 16 people,
divide each piece
you divide them
smaller because you
of what
Apple pie
Divide
16 hungry people

In this phase, you will be exploring patterns, squares, cubes, and roots. Testing rules for special cases will help you write your own mathematical arguments that apply to more general cases in mathematics.

Knowing how to represent squared and cubed numbers with drawings and cubes will help you understand more about them. How do you think you could show $4^3$?

# Roots, Rules, and Arguments

## WHAT'S THE MATH?

*Investigations in this section focus on:*

### NUMBER and RELATIONSHIPS

- Understanding how to find the square root, cube root, square, and cube of a number
- Investigating relationships among 0, 1, proper fractions, and negative numbers
- Understanding how to raise a number to an exponent

### MATHEMATICAL REASONING

- Making mathematical arguments
- Evaluating mathematical arguments others have written

### PATTERNS and FUNCTIONS

- Describing and finding perfect squares in tables and rules

# 5 Perfect Pattern Predictions

EXPLORING PATTERNS IN PERFECT SQUARES

**Any number multiplied by itself is called the square of that number.** When the number that is multiplied by itself is an integer, the result is called a perfect square. In the last phase, you found rules for signed number operations. Can you find a rule to describe a pattern in perfect squares?

## Investigate Perfect Squares

**What is the pattern in the increase from one perfect square to another?**

Look at the perfect squares shown in the box What's the Pattern? The number that is multiplied by itself to produce each square is called the square root. For example, the square root of 9 is 3. We can use a radical sign to write this as $\sqrt{9} = 3$.

1. The box What's the Pattern? shows the perfect squares that result when you square each whole number from 1 through 5. By how much is each perfect square increasing sequentially?
2. Make a table showing the increase when each whole number from 1 through 12 is squared. Label your table with three columns: Original Number, Perfect Square, and Increased By.
3. Extend your table to find between which two perfect squares the increase will be 21, 27, and 35.

**What's the Pattern?**

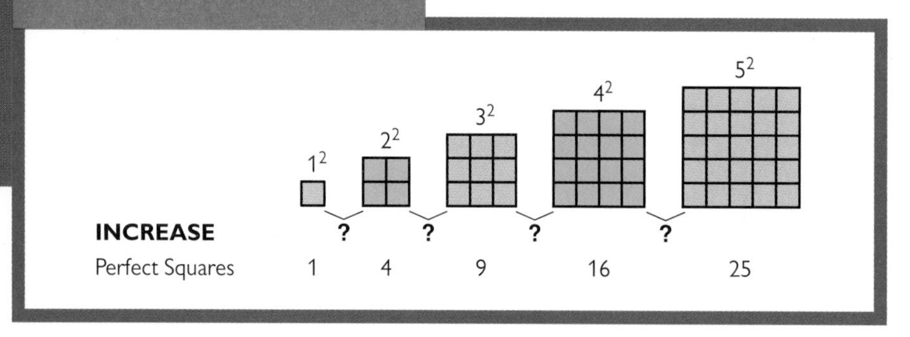

## Find a Method to Predict the Increase

Come up with a method for figuring out the increase between any two perfect squares. You may find it helpful to use a calculator to investigate squares and square roots in coming up with a method.

**1** Use words, diagrams, or equations to explain your method.

**2** Think about how each step in your method relates to the way the perfect squares are shown with cubes. What do the numbers in your method represent in the cubes?

**3** Use your method to figure out the increase between each of the following perfect squares and the perfect square that comes next: 529; 1,089; 2,401.

**How can you predict the increase between any two perfect squares?**

## Write a Rule for the Pattern

Think about the method you came up with for predicting the increase for any perfect square.

- Write a rule that describes how to predict what the increase between perfect squares will be.
- Every positive square number has a positive and a negative square root. For example, 36 has square roots of 6 and $-6$ since $6^2 = 36$ and $(-6)^2 = 36$. $-6$ is the negative square root of 36. Does the rule you wrote work for the negative square root? If so, give examples. If not, give counterexamples and revise your rule.

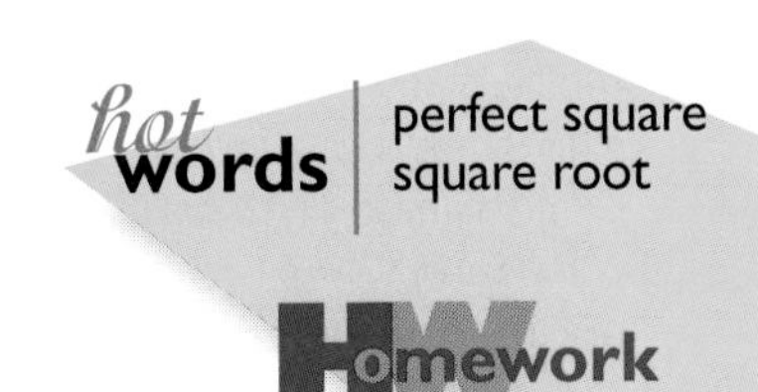

# Counterexamples and Special Cases

FORMULATING A MATHEMATICAL ARGUMENT

**In Lesson 5, you investigated patterns in perfect squares when positive and negative integers are squared.** In this lesson, you will examine a mathematical argument about the result of squaring any number, looking for counterexamples and checking for special types of numbers. This will help you think of ways to revise the mathematical argument so that it is always true.

## Find Counterexamples to Dan's Rule

**How can you examine a mathematical argument to find counterexamples?**

After you have read Dan's Mathematical Argument, work with your group to try to find counterexamples to Dan's rule. Work through some or all of these questions:

1. What do you think of Dan's rule?
2. Can you find any specific counterexamples for Dan's rule?
3. Can you find any types of numbers for which every number forms a counterexample?
4. What do you think is the smallest number for which Dan's rule is true? the largest number?

### Dan's Mathematical Argument

**What is my rule?** My rule states: "The square of any number is always larger than the original number." For example, if I start with the number 8 and square it, I get 64. 64 is definitely larger than 8.

**How did I figure out my rule?** I started by choosing different numbers, like 3, 8, 12, 47 and 146. Each time I squared them, I got a larger number. I saw that as the original numbers got larger, the square numbers also got larger very quickly.

**For what special cases is my rule true?** I tried small and large numbers, like 3 and 146, and my rule was always true.

| Original Number | Square Number |
|---|---|
| 3 | 9 |
| 8 | 64 |
| 12 | 144 |
| 47 | 2,209 |
| 146 | 21,316 |

## Revise a Mathematical Argument

How can you make a mathematical argument that is always true?

Think about Dan's Mathematical Argument and read it again if necessary. Write a new version of the argument that is correct and complete. Use the different points below as guidelines to make sure you have included all the information necessary to make your argument a strong one. Show all of your work.

**1** What is your rule?

a. Did you state your rule clearly?

b. Could someone who had not already done the investigation understand your rule?

c. Did you describe your rule generally so that it can apply to more than just a few numbers?

**2** How did you figure out your rule?

a. What methods did you use to figure out your rule?

b. What counterexamples did you find where your rule did not work?

**3** Does your rule apply to special cases?

a. Does your rule work for 0? for 1? for fractions? for negative numbers?

b. Are there other cases for which your rule works or does not work?

c. If your rule does not work for some cases, explain why it doesn't.

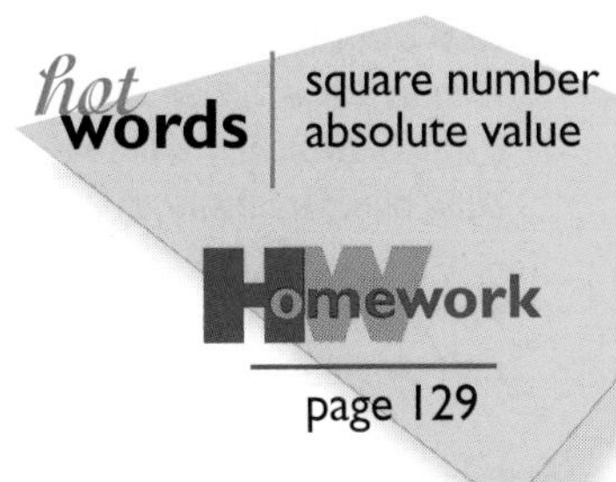

# 7 Root Relationships

WRITING AND REVISING A MATHEMATICAL ARGUMENT

**You have already learned what a square root is and investigated mathematical arguments involving square numbers and square roots.** In this lesson, you will explore cubes of numbers and cube roots. A perfect cube is a number that results when you use an integer as a factor three times.

## Verify Dan's Mathematical Argument

**What is missing from Dan's mathematical argument?**

Follow the steps below to verify if Dan's argument is always true.

1. Check Dan's argument for counterexamples. List them and show why they do not fit Dan's rule.
2. Check Dan's argument for special cases. Is his rule true for 0, 1, fractions, and negative numbers? For each case, show the original number (or numbers) and describe how it compares to the cube of the number.
3. Do you have any special cases of your own that you want to check? Try them out and describe what you find.

### Dan's 2nd Mathematical Argument

**What is my rule?** My rule states: "The cube of any number is always larger than the original number." For example, if I start with the number 3 and cube it, I get 27 and 27 is definitely larger than 3.

**How did I figure out my rule?** I started by choosing different numbers, like 2, 2.2, 3, 3.5, and 4. Each time I cubed them, I got a larger number. I saw that as the original numbers got larger, the cube numbers also got larger.

**For what special cases is my rule true?** I tried whole numbers and decimals, like 2 and 2.2, and my rule was always true.

| Original Number | Square Number |
|---|---|
| 2 | 8 |
| 2.2 | 10.648 |
| 3 | 27 |
| 3.5 | 42.875 |
| 4 | 64 |

## Look at Square and Cube Roots

How do positive and negative numbers relate to cube and square roots?

As you answer the questions below, think about rules you might make.

**1** Can you find each of these numbers? If you can, give an example. If you cannot, write "no."

a. a positive number with a positive cube root
b. a positive number with a positive square root
c. a positive number with a negative cube root
d. a positive number with a negative square root
e. a negative number with a positive cube root
f. a negative number with a positive square root
g. a negative number with a negative cube root
h. a negative number with a negative square root

**2** What rules can you come up with for square and cube roots and positive and negative numbers? Why do your rules work?

## Write and Revise a Mathematical Argument

Follow these steps to write and revise your own mathematical argument about cubes of numbers and original numbers.

**1** Write a mathematical argument that includes a statement of the basic rule or mathematical idea, the methods used to figure out the rule, and any counterexamples you found where your rule did not work. Be sure to include a description of what happens in special cases.

**2** Trade mathematical arguments with a partner. Look at the handout Guidelines for Writing Your Own Mathematical Argument.

a. Did your partner meet all of the guidelines?

b. Is there information in your partner's argument that you don't understand?

c. Is there information you think your partner left out?

**3** Revise your mathematical argument using your partner's feedback and any other information you have learned.

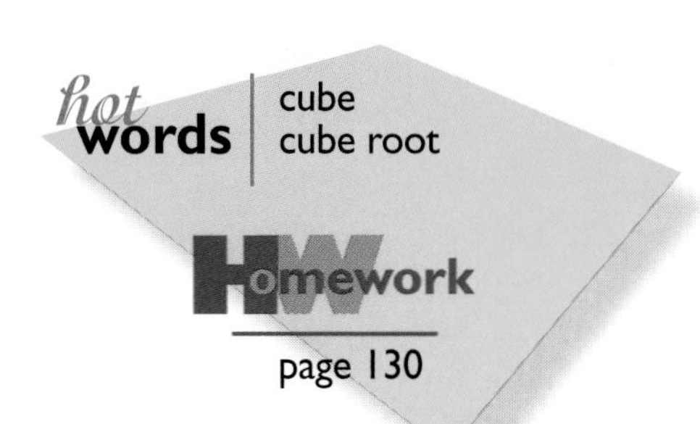

# 8 A Powerful Argument

WRITING A MATHEMATICAL ARGUMENT ABOUT A PATTERN

**In this lesson, you will search for perfect squares in a chart which has numbers raised from the power of 2 to the power of 6.** Exploring patterns in the location of these perfect squares helps you understand more about squares, roots, and exponents. This will prepare you for writing your own mathematical argument about one of the patterns.

What patterns of perfect squares can you find in The Powers Chart?

## Create a Powers Chart

Make a chart like the one shown and follow the next three steps.

**The Powers Chart**

| | $\square^1$ | $\square^2$ | $\square^3$ | $\square^4$ | $\square^5$ | $\square^6$ |
|---|---|---|---|---|---|---|
| **1** | 1 | | | | | |
| **2** | 2 | | | | | |
| **3** | 3 | | | | | |
| **4** | 4 | | | | | |
| **5** | 5 | | | | | |
| **6** | 6 | | | | | |
| **7** | 7 | | | | | |
| **8** | 8 | | | | | |
| **9** | 9 | | | | | |

**1** Use your calculator to fill in the chart by raising each number to the exponent shown at the top of the column.

**2** When you have filled in all the numbers, use your calculator to check which numbers are perfect squares and circle them.

**3** Do you see any patterns in the rows and columns where numbers are circled?

## Write a Mathematical Argument

We have talked about a rule for the pattern found in the rows of the Powers Chart. This statement describes the pattern found in the columns of the chart: "If you raise any number to an even power, the result will be a perfect square."

**1** Test the statement above and revise it by using the handout Guidelines for Writing Your Own Mathematical Argument.

**2** Remember to include a statement of the basic rule, the methods used to figure out the rule and any counterexamples, and a description of what happens with special cases.

**How would you write a mathematical argument about one of the patterns?**

## Test and Revise a Mathematical Argument

Work with a partner who will provide feedback and revise your mathematical argument by completing the following:

- Trade mathematical arguments with a partner.
- Read your partner's argument, checking to be sure your partner responded well to all of the questions in the Guidelines for Writing Your Own Mathematical Argument. Look for anything you don't understand and or anything your partner might have left out, and write down your comments.
- Return the mathematical argument and comments to your partner.
- Make revisions to your mathematical argument based on the written comments you receive from your partner.

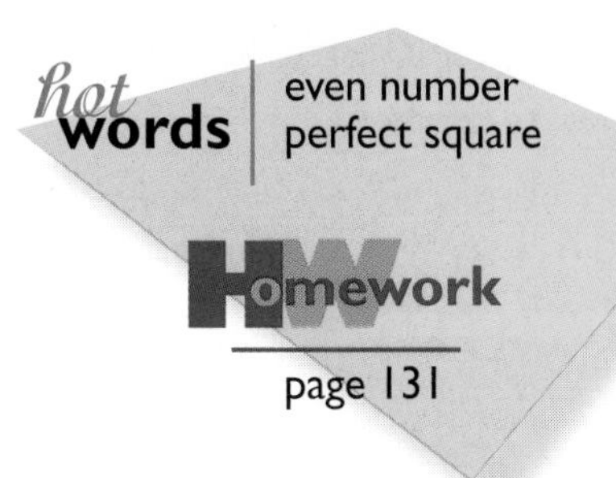

# PHASE THREE

In this phase, you will write mathematical arguments about patterns involving divisibility, prime numbers, and factors. These topics have interested mathematicians since ancient times. A method for finding prime numbers was developed by a Greek astronomer around 240 B.C.

Today, powerful computers are used to search for prime numbers and prime numbers are used in computer codes for security.

# Primes, Patterns, and Generalizations

## WHAT'S THE MATH?

*Investigations in this section focus on:*

### NUMBER THEORY

- Understanding divisibility, primes, factors, and multiples

### MATHEMATICAL REASONING

- Making and understanding mathematical arguments
- Evaluating mathematical arguments others have written

### PATTERNS and FUNCTIONS

- Using patterns to describe general rules
- Finding patterns in factors, primes, squares, or cubes

# Three-Stack Shape Sums

INVESTIGATING A PATTERN IN DIVISIBILITY BY 3

**In the last phase, you used cubes to model square numbers. In this lesson, you model numbers by organizing cubes into stacks of three.** Looking at numbers modeled in this way will help you investigate the statement: "The sum of any three consecutive whole numbers will always be divisible by 3."

## Create Numbers Using the 3-Stacks Model

**What patterns can you see when you make numbers with the 3-stacks model?**

We say that one number is divisible by a second number if the first number can be divided evenly by the second number, leaving a remainder of 0. With the class, you have tried to see if you could find any three consecutive whole numbers whose sum is not evenly divisible by 3. Work with a partner and follow the investigation steps below to explore some patterns in consecutive numbers.

1. Use cubes to make each number from 1 to 15 following the 3-stacks model. On grid paper, record each number like the numbers shown in How to Use the 3-Stacks Model.
2. Look for patterns in the way the numbers look in the 3-stacks model. Be ready to describe any patterns you notice.

### How to Use the 3-Stacks Model

| L Number | b Number | Rectangle Number |
|---|---|---|
| 4 | 5 | 6 |

Each time you have a stack of three cubes, start on a new column to the right.

## Investigate the Sum of Numbers

How can you predict whether the sum of two numbers will be an L number, a b number, or a rectangle number?

In the last investigation, you found that numbers you made with cubes in the 3-stacks model could be described as L numbers, b numbers, or rectangle numbers. Follow the steps below to think about what it means to add any two kinds of numbers.

1. Make a table like the one shown. Put all the possible combinations of L numbers, b numbers, and rectangle numbers you can think of in the first two columns.
2. Then think about what your result would be if you added these kinds of numbers. Put that information in the third column.
3. What conclusions can you make about the sum of two numbers by looking at your table?

| If you add this kind of number: | to this kind of number: | you get this kind of number: |
|---|---|---|
| | | |
| | | |
| | | |
| | | |

## Write and Revise a Mathematical Argument

Through class discussion, you have learned that a whole number is divisible by its factors. Write your own mathematical argument for this statement: "Whenever you add three consecutive whole numbers together, the sum will be divisible by 3."

- Remember to check your thinking. Are there any special cases you should consider? What happens with each of the special cases?
- Trade your argument with a partner and ask your partner to comment on it.
- Revise your mathematical argument based on your partner's comments.

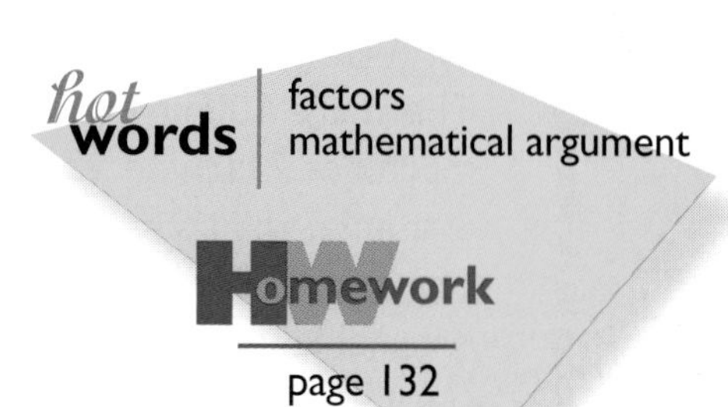

# 10 A Stretching Problem

GENERALIZING ABOUT FACTORS, MULTIPLES, AND PRIMES

**A prime number is a number that has exactly two factors, 1 and itself.** In this lesson, you will be looking at patterns in prime numbers to solve a problem at a bubble gum factory. Then you will write your own mathematical argument about prime numbers.

How can factors and multiples help you think about prime numbers?

## Find the Unnecessary Machines

First, read the information on this page about the Bubble Gum Factory to understand how it operates. Then read the handout The Bubble Gum Factory Script to find out what problem you can solve for the Bubble Gum Factory.

1. Look at the handout The Unnecessary Machines. Each one of the squares on The Unnecessary Machines is one of the machines in the Bubble Gum Factory. Some of the machines are unnecessary because combinations of other machines could be used instead.

2. Figure out which machines are actually unnecessary and cross them off. Be prepared to discuss with the class why you crossed off these machines.

### The Bubble Gum Factory

At the Bubble Gum Factory, 1-inch lengths of gum are stretched to lengths from 1 inch to 100 inches by putting them through a stretching machine. There are 100 stretching machines. Machine 23, for example, will stretch a piece of gum to 23 times its original length.

## Investigate the Necessary Machines

What prime numbers can you use to make other numbers?

Use the questions below to find out how you could combine the *necessary* machines to get other lengths. Any of these machines may be used more than once to give the requested length.

1. What machines could you use to get the lengths: 15? 28? 36? 65? 84?
2. For each of the lengths above, what other machines could have been used that were unnecessary?
3. Which lengths between 1 and 100 would come out if the bubble gum went through five machines and all 5 machines were necessary ones?
4. Which length between 1 and 100 requires the greatest number of necessary machines? How did you figure out your answer?

## Write and Revise a Mathematical Argument

Write your own mathematical argument about this statement: "Any number can be written as the product of prime factors."

- Consider all the special cases we have used in this unit. Think about whether the special cases should be included in your argument. If a special case does not apply, it is sufficient to say so in your mathematical argument.
- Look for special cases other than 0, 1, proper fractions, and negative numbers. Your mathematical argument should describe the range of numbers for which the rule is true.
- Work with a partner to read and comment on each other's arguments.
- Revise your mathematical argument based on your partner's comments.

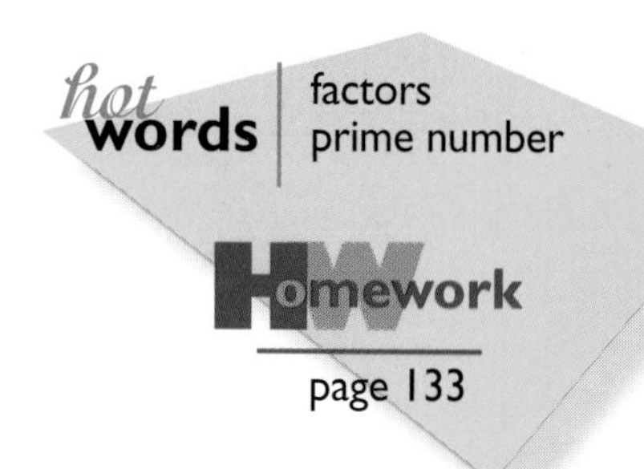

# 11 Pattern Appearances

GENERALIZING ABOUT PATTERNS IN A MULTIPLICATION CHART

**You will continue to think about squares, cubes, primes, and factors as you look for patterns in the Multiplication Chart.** From these patterns you can make some general rules. Your goal will be to find out how many times any number would appear in the Multiplication Chart if the chart continued into infinity!

## Find Out How Often Numbers 11–25 Appear

**How many times will 11–25 appear on the Multiplication Chart?**

After you work with the class to find out how many times the numbers 1–10 appear on the handout Multiplication Chart, expand your investigation by responding to the directions below. You will need the table you made with the class for this investigation.

1. How many times will 11 appear on the chart? 12? 13? Figure out how many times each of the numbers from 11 to 25 would appear on the chart and add them to your table. Remember to include how many times the number would appear if the chart went on into infinity, not just the number of times it appears on the Multiplication Chart you have.

2. How might you predict the number of times any number will appear as a product on the chart?

**Multiplication Chart**

| × | 1 | 2 | 3 | 4 | 5 | 6 | 7 | 8 | 9 | 10 | 1 |
|---|---|---|---|---|---|---|---|---|---|---|---|
| **1** | 1 | 2 | 3 | 4 | 5 | 6 | 7 | 8 | 9 | 10 | 1 |
| **2** | 2 | 4 | 6 | 8 | 10 | 12 | 14 | 16 | 18 | 20 | 2 |
| **3** | 3 | 6 | 9 | 12 | 15 | 18 | 21 | 24 | 27 | 30 | 3 |
| **4** | 4 | 8 | 12 | 16 | 20 | 24 | 28 | 32 | 36 | 40 | 4 |
| **5** | 5 | 10 | 15 | 20 | 25 | 30 | 35 | 40 | 45 | 50 | 5 |
| **6** | 6 | 12 | 18 | 24 | 30 | 36 | 42 | 48 | 54 | 60 | 6 |

## Predict in Which Column a Number Belongs

How can you predict which column of your table a number belongs in?

You will need the table you made for this investigation. After exploring with the class which columns of your table the numbers 27, 36, 37, and 42 belong in, answer the questions below.

**1** Choose five other numbers of your own between 26 and 100. What columns of your table do you think they belong in? Use the Multiplication Chart to check your predictions and then write the numbers in the appropriate columns on your table.

a. How would you describe the numbers that appear in the 2-times column of your table?

b. How would you describe the numbers that appear in the 3-times column of your table?

c. Choose one other column of your table. How would you describe the numbers that appear in that column?

**2** Identify numbers that belong in particular columns of your table.

a. Can you think of a number that belongs in the 2-times column that is not already there? Write it in the column on your table.

b. Can you think of a number that belongs in the 3-times column that is not already there? Write it in the column on your table.

c. Can you think of a number that belongs in the column you chose in item **1c** that is not already there? Write it in the column on your table.

## Generalize About the Column in Which Any Number Belongs

After the class discussion, write an answer to this question: How could you predict which column of your table *any* number would belong in?

- Be sure to include your own thinking about the question.
- Use examples to explain your answer.

# 12 The Final Arguments

WRITING A COMPLETE MATHEMATICAL ARGUMENT

**You will use what you have learned in this unit to write an argument about the numbers in the Multiplication Chart.** You will also learn from others as you work together in small groups to share ideas about mathematical arguments.

## Share Ideas About Mathematical Arguments

What will you write your mathematical argument about?

Choose something to write your mathematical argument about from the table you made for the handout Multiplication Chart. It could be one of the columns of your table, such as the 2× column, or it could be the numbers that appear most frequently. Your group may choose the same thing, or each person in the group could choose something different.

1. What is your rule for what you chose? How can you state your rule so that anyone who has not done the investigation yet will know what you are talking about?
2. Can you find any counterexamples to your rule? If so, how will you change your rule to take them into account?
3. Does your rule work for special cases? Why or why not?

### Characteristics of a Good Mathematical Argument

A good mathematical argument should include the following:

- a rule that is general and clearly stated
- a description of how the rule was figured out, including a search for counterexamples (The rule should be revised if any counterexamples are found.)
- a description of special cases to which the rule applies

## Write Your Own Mathematical Argument

Based on the discussion you had in your group, write your own mathematical argument using the following information:

1. Include all of the Characteristics of a Good Mathematical Argument on page 122.
2. Review the table you created and the Multiplication Chart from Lesson 11. Refer to the writing you did in Lesson 11 about how to predict which column of your table *any* number on the Multiplication Chart would belong in.

**What do you include in a well-written mathematical argument?**

## Share and Revise Your Mathematical Argument

When you have completed your mathematical argument, trade with a partner.

- As you read your partner's argument, think of yourself as a teacher looking for a well-written mathematical argument. Write comments that you think will help your partner improve the argument.
- Use your partner's comments to revise your mathematical argument. Review the Characteristics of a Good Mathematical Argument again to make sure you have included all that is required.

# Statements About Signs

## Applying Skills

**For items 1–8, a pink square represents a positive cube and a green square represents a negative cube.**

**1.** What number do these cubes show?

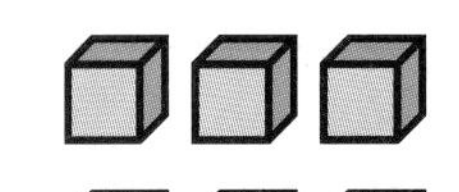

**2.** What number do these cubes show?

**3.** What number do these cubes show?

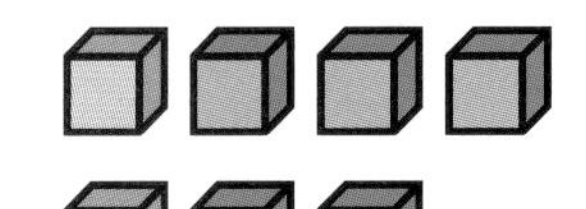

**For items 4–8, use the pictures of cubes shown to solve a subtraction problem.**

**4.** What number do the cubes show?

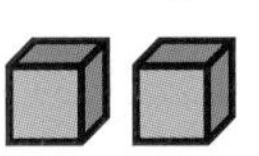

**5.** What has been added?

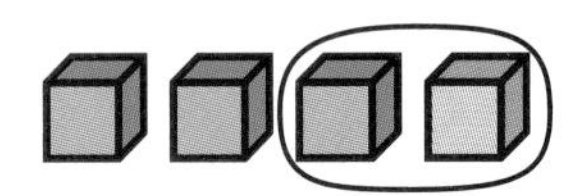

**6.** What has been removed?

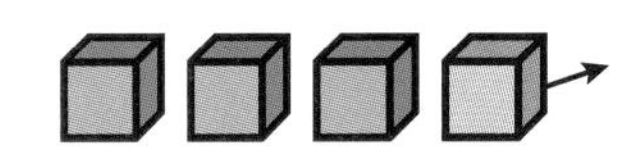

**7.** What number will be shown by the remaining cubes?

**8.** What was the subtraction problem? What is the answer?

**For items 9–13, solve each problem by drawing pictures of cubes.**

**9.** $3 + (-1)$ **10.** $5 - (-3)$

**11.** $-2 + (-3)$ **12.** $-4 - (-1)$

**13.** $-6 - 4$

## Extending Concepts

**14.** Make up a subtraction problem with a positive answer which can be solved using cubes, and for which you need to add some zero-pairs to your cubes. Draw the cubes to illustrate the problem.

**For items 15–16, tell whether each statement is always true. If it is not always true, find a counterexample. Then rewrite the rule so that it is always true.**

**15.** If you add a positive proper fraction and a negative proper fraction, you will always get a number smaller than 1.

**16.** If you subtract a positive fraction from a positive fraction, you will always get a number smaller than $\frac{3}{4}$.

## Writing

**17.** Answer the letter to Dr. Math.

Dear Dr. Math,
My theory was: "The sum of a positive number and a negative number is always positive." I found 27 examples that worked. My friend Kate found one counterexample. So I figured 27 to 1, my theory must be good. But Kate said she only needed one counterexample to disprove my theory. Is this true?
Positively Exhausted

# Counterexamples and Cube Combinations

## Applying Skills

**For items 1–8, solve each problem and write the entire equation.**

**1.** $3 + (-2)$ **2.** $7 - (-5)$ **3.** $-8 + (-2)$

**4.** $-3 - 8$ **5.** $4 + (-2) + (-3)$

**6.** $-5 + 3 - 6$ **7.** $6 - 2 - (-1)$

**8.** $4 + (-2) - 3 + (-9)$

**9.** Each number in the tables below is found by subtracting the number in the top row from the number in the leftmost column. Copy each puzzle and fill in all the missing values.

| − | −2 | 3 | −4 | 8 |
|---|---|---|---|---|
| 1 | | | | |
| 5 | | 2 | | |
| −6 | | | | |
| 0 | | | | |

| − | −3 | 4 | | |
|---|---|---|---|---|
| −2 | | | | |
| 7 | | 3 | 5 | |
| | | | | −3 |
| | | | 0 | 3 |

## Extending Concepts

**10.** What combinations of four cubes can you come up with? For each combination, make up an addition or subtraction problem for which the answer is $-4$.

**11.** Find three different combinations of cubes, each representing the number $-7$. For each combination, make up a problem for which the answer is 4.

**12.** If you add a positive number and a negative number, how can you tell whether the answer will be positive, negative, or zero?

**13.** Suppose you start with the number $-3$ and add a positive number. What can you say about the positive number if the answer is positive? negative? zero?

**14.** If you add two positive numbers and one negative number, will the answer always be positive? If not, how can you tell whether the answer will be positive, negative, or zero?

## Making Connections

**The *emu* is a large flightless bird of Australia. The *paradoxical frog* of South America is so-called because the adult frog is smaller than the tadpole. Tell whether each statement in 15–19 is always true. If the statement is not always true, give a counterexample.**

**15.** Some birds can't fly.

**16.** No bird can fly.

**17.** All birds can fly.

**18.** In every species, the adult is bigger than the young.

**19.** There is no species in which the adult is bigger than the young.

# More Cases to Consider

## Applying Skills

The multiplication 3 × 2 can be shown as 3 groups of 2.

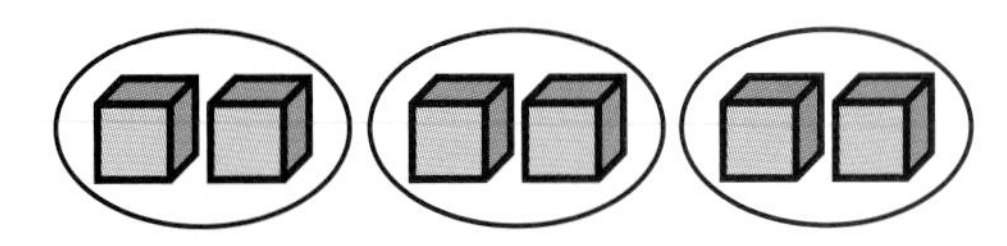

Draw cubes similar to the ones above to show each multiplication in items 1–3. Use shaded squares to represent negative numbers.

**1.** $2 \times 4$ **2.** $3 \times (-5)$ **3.** $4 \times 3$

Solve each problem in items 4–10.

**4.** $5 \times (-8)$

**5.** $6 \div (-2)$

**6.** $-9 \div (-3)$

**7.** $-8 \times (-7)$

**8.** $6 \times (-2) \div 3$

**9.** $-5 \times 4 \div (-8)$

**10.** $-4 \div 4 \times (-10) \div (-5)$

## Extending Concepts

In items 11 and 12, look for different paths in the puzzles that equal some result. For each path you find, write an equation. You may move in any direction along the dotted lines, but you may only use each number once. Your path must start in the top row and end in the bottom row.

**11.** You may use only multiplication on this puzzle. A path that equals 24 is shown here.

**Multiplication Puzzle**

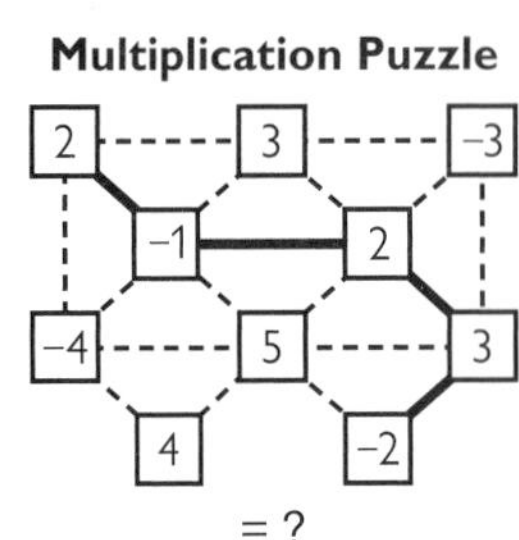


**a.** What path can you find that equals −240?

**b.** What is the longest path you can find? What does it equal?

**12.** In the puzzle below, you may use either multiplication or division. At each step, you choose which operation you want to use. You must use each operation at least once.

**Multiplication and Division Puzzle**

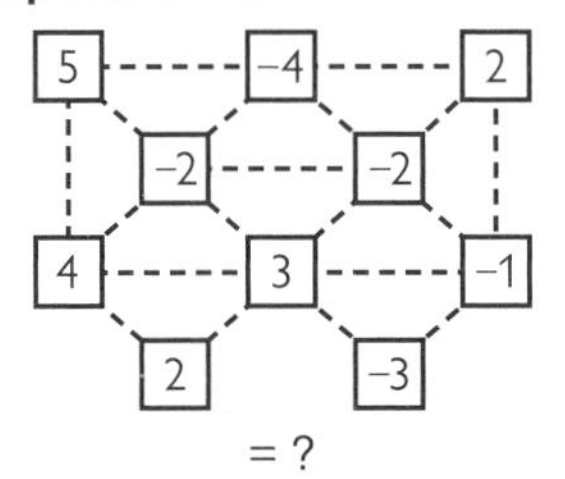


**a.** What path can you find that gives you an answer between −1 and −2?

**b.** What path can you find that gives you a positive answer less than 1?

## Writing

**13.** Make up your own puzzle that uses multiplication, division, addition, and subtraction. Write two questions to go with your puzzle and make an answer key for each question.

# Rules to Operate By

## Applying Skills

**For each subtraction problem in items 1–6, write an equivalent addition problem. For each addition problem, write an equivalent subtraction problem.**

**1.** $8 - (-6)$ **2.** $9 - 2$ **3.** $7 + (-11)$

**4.** $10 - (-1)$ **5.** $3 + (-12)$ **6.** $4 + (-6)$

**7.** Using any pair of the numbers 5, $-5$, 8, and $-8$, write four different addition or subtraction problems that have the answer $-13$.

**For items 8–9, use $-6$, $-1$, 2, $-9$, addition, and subtraction to solve each problem.**

**8.** What problem can you find with the least possible answer?

**9.** What problem can you find for which the answer is zero?

**10.** Using the numbers $-6$, $\frac{1}{2}$, $-2$, and 4, and any three operations, what problem can you find for which the answer lies between $-2$ and 0?

## Extending Concepts

**To answer items 11–12 use the puzzle shown. You may use addition, subtraction, multiplication, or division. You must use each operation at least once, and your path must start in the top row and end in the bottom row. Remember, you may move in any direction along the dotted lines, but you may only use each number once.**

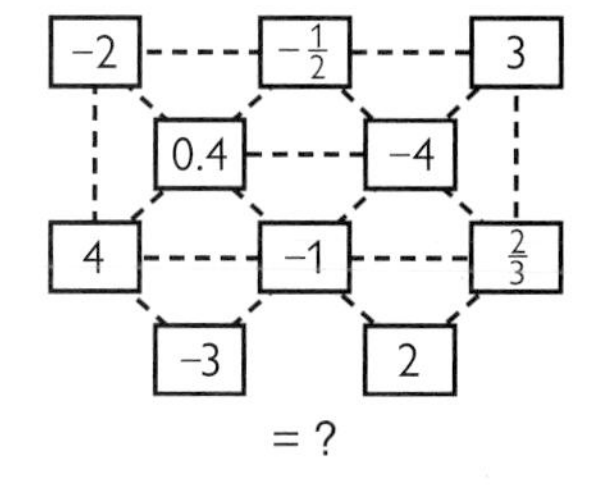


**11.** What path gives an answer between 1 and 2?

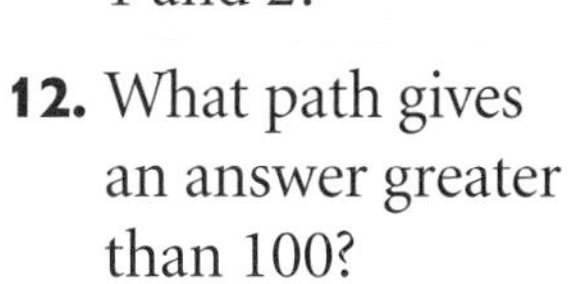

**12.** What path gives an answer greater than 100?

**13.** What statement could you write for which $2 + (-6) = -4$ is the counterexample?

**Tell whether each statement in items 14–15 is always true or not always true. If it is not always true, find a counterexample.**

**14.** If you subtract a negative fraction from a positive fraction, you will always get a number greater than $\frac{1}{4}$.

**15.** The product of two numbers will always be greater than the sum of the same two numbers.

## Making Connections

**For items 16 and 17, use the following information:**

The Dead Sea is a salt lake lying on the Israel-Jordan border. At 1,292 feet below sea level, its surface is the lowest point on earth. At 29,028 feet, the top of Mount Everest is the highest point on earth.

**16.** Write a subtraction equation to find the elevation difference between the top of Mount Everest and the surface of the Dead Sea.

**17.** Write an equivalent addition equation.

# Perfect Pattern Predictions

## Applying Skills

**For items 1–4, write each square using exponents and find their values.**

**1.** 7 squared **2.** 12 squared

**3.** 16 squared **4.** 29 squared

**For items 5–10, find each square root.**

**5.** $\sqrt{49}$ **6.** $\sqrt{81}$ **7.** $\sqrt{196}$

**8.** $\sqrt{484}$ **9.** $\sqrt{1{,}024}$ **10.** $\sqrt{1{,}444}$

**For items 11–14, find the increase between each pair of perfect squares.**

**11.** 5th and 6th perfect squares

**12.** 11th and 12th perfect squares

**13.** 18th and 19th perfect squares

**14.** 30th and 31st perfect squares

**For items 15–20, between which two perfect squares will the increase be:**

**15.** 13? **16.** 19? **17.** 31?

**18.** 109? **19.** 363? **20.** 37?

**For items 27–29, use the pictures shown to answer the questions.**

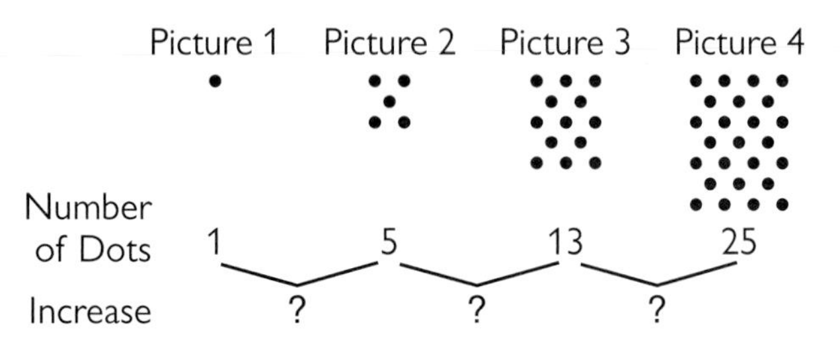


**27.** Find the increase in the number of dots between each of the figures shown. What pattern do you notice? What will be the increase in the number of dots between the 9th and 10th figures?

**28.** Describe a rule you could use to find the increase in the number of dots between any two figures. What will the increase be between the 100th and 101st figures?

**29.** The number of dots in the third figure is $3^2 + 2^2$ or 13. The number of dots in the 4th figure is $4^2 + 3^2$ or 25. What pattern do you notice? How many dots will be in the 40th figure?

## Extending Concepts

**For items 21–26, find each square root.**

**21.** $-\sqrt{36}$ **22.** $-\sqrt{64}$ **23.** $-\sqrt{144}$

**24.** $-\sqrt{324}$ **25.** $-\sqrt{1{,}089}$ **26.** $-\sqrt{1{,}764}$

## Making Connections

**30.** The pyramid of Khufu in Egypt was built in 2680 B.C. as a burial tomb for the king. It has a square base measuring 756 feet on each side. The formula for the area of a square is $s^2$, where $s$ is the length of a side. If each side of the pyramid was one foot longer, how much greater would the area of the base be?

# Counterexamples and Special Cases

## Applying Skills

For items 1–6, tell whether the square of each number is greater or less than the number itself. Do not calculate the square.

**1.** 5 **2.** 0.9 **3.** $\frac{1}{5}$ **4.** $-18$ **5.** $\frac{2}{3}$ **6.** $-0.1$

For items 7–9, show each multiplication by drawing a sketch, then give the result.

**7.** $\frac{1}{3} \times \frac{1}{2}$ **8.** $\frac{1}{2} \times \frac{1}{2}$ **9.** $2 \times \frac{1}{3}$

**10.** Copy and fill in a table like the one shown. Estimate the square root of each number in the table. Write down your estimate and then test it by squaring it. Repeat this process two more times. Then use your calculator to calculate how much your best estimate differed from the answer shown on the calculator.

| | What is the square root of: | Estimate 1 | Estimate 2 | Estimate 3 | By how much did your best estimate differ? |
|---|---|---|---|---|---|
| **a.** | 43 | | | | |
| **b.** | 86 | | | | |
| **c.** | 306 | | | | |
| **d.** | 829 | | | | |

## Extending Concepts

For items 11–14, consider the following rule: "The square root of a number is always less than the original number."

**11.** Give three counterexamples to the rule.

**12.** Test whether the rule works for each of these types of numbers: negative numbers, proper fractions, 0, 1, and positive numbers greater than 1. For types of numbers where the rule does not work, explain why.

**13.** Write a correct version of the rule.

**14.** Why do you think that 0 and 1 are often tested as special cases?

## Writing

**15.** You learned in class that the square of a proper fraction is less than the original fraction. Use this theory to explain the meaning of counterexamples. You may want to give examples in your explanation.

# Root Relationships

## Applying Skills

Calculate items 1–6. To figure out the cube roots, you may want to use a calculator to guess and check.

**1.** $3^3$ **2.** $11^3$ **3.** $(-4)^3$

**4.** $(-9)^3$ **5.** the cube root of 125

**6.** the cube root of $-1{,}000$

For items 7–12, tell whether each number is a perfect cube.

**7.** 64 **8.** 16 **9.** $-8$

**10.** 25 **11.** 343 **12.** 1,728

For items 13–15, consider the following rule: "The cube of any number is greater than the square of the same number."

**13.** Give three counterexamples to the rule.

**14.** Test whether the rule works for each of these special cases: proper fractions, 0, 1, negative numbers, and positive numbers greater than 1. If there are some special cases for which the rule does not work, explain why not.

**15.** Write a new correct version of the rule.

## Extending Concepts

**16.** Copy and fill in a table like the one shown. For each cube root, think about the two whole-number cube roots it might lie between. Write your answer in the second column. Estimate the cube root to the nearest hundredth. Write your answer in the third column.

| | What is the cube root of: | Lies between cube roots: | Estimate |
|---|---|---|---|
| **a.** | 50 | | |
| **b.** | 200 | | |
| **c.** | 520 | | |

**17.** Is it possible to find a number with a negative cube root and a positive square root? If so, give an example. If not, explain why not.

## Making Connections

For item 18, use the following:

Kepler's third law states that for all planets orbiting the sun, the cube of the average distance to the sun divided by the square of the period (the time to complete one revolution around the sun) is about the same.

**18.** Look at the table shown. Here is an example of the calculations for Earth:

$d^3 = 93^3 = 804{,}357$

$T^2 = 365^2 = 133{,}225$

$$\frac{d^3}{T^2} = \frac{804{,}357}{133{,}225} = 6.04$$

Is the law true for Earth, Venus, and Mercury?

| Planet | Average distance to sun in millions of miles (d) | Period in days (T) |
|---|---|---|
| Earth | 93 | 365 |
| Venus | 67 | 225 |
| Mercury | 36 | 88 |

# A Powerful Argument

## Applying Skills

**Calculate items 1–10. To figure out the cube roots, you may want to use the calculator to guess and check.**

**1.** $11^4$  **2.** $7^3$  **3.** $8^6$

**4.** $4^5$  **5.** $3^7$  **6.** $\sqrt{529}$

**7.** the cube root of 729

**8.** $\sqrt{289}$  **9.** $\sqrt{1,225}$

**10.** the cube root of $-1331$

**Without using a calculator, identify each number in items 11–19 that you know is a perfect square. Tell why you know.**

**11.** $4^5$  **12.** $5^6$  **13.** $6^3$

**14.** $8^7$  **15.** $2^{10}$  **16.** $3^9$

**17.** $9^3$  **18.** $7^8$  **19.** $11^4$

**20.** Use your calculator to calculate each power in items 11–19.

**21.** Use the square root key on your calculator to find the square root for each number in items 11–19. Identify which of the numbers is a perfect square.

## Extending Concepts

**22.** Copy the table shown. Which rows and columns in the table do you think are perfect cubes? How can you tell?

**The Powers Chart**

| | $\square^1$ | $\square^2$ | $\square^3$ | $\square^4$ | $\square^5$ | $\square^6$ |
|---|---|---|---|---|---|---|
| **6** | | | | | | |
| **7** | | | | | | |
| **8** | | | | | | |

**23.** Complete the table by raising each number to the exponent at the top of the column.

**24.** Use guess-and-check and your calculator to check which numbers in the table are perfect cubes and circle them.

**25.** Someone made this mathematical argument: $x^a \times x^b = x^{a+b}$. What do you think about it?

## Making Connections

**Use this information for items 26 and 27.**

Jainism is a religious system of India which arose in the 6th century B.C. and is practiced today by about 2 million people. According to Jaina cosmology, the population of the world is a number which can be divided by two 96 times. This number can be written in exponent form as $2^{96}$.

**26.** Is $2^{96}$ a perfect square? How can you tell?

**27.** Is $2^{96}$ a perfect cube? How can you tell?

# Three-Stack Shape Sums

### Applying Skills

In the 3-stacks model, cubes are stacked in columns with a height of 3 cubes. The rightmost column may contain 1 cube (L numbers), 2 cubes (b numbers), or 3 cubes (rectangle numbers).

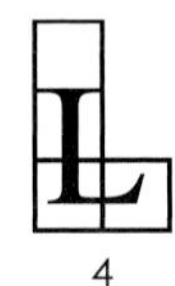
4

5

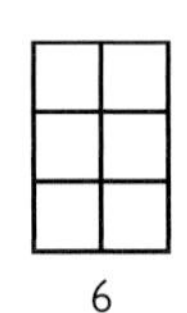
6

**For items 1–6, draw a sketch showing how you could make each number using the 3-stacks model. Tell which kind of number each number is: a rectangle number, a b number, or an L number.**

**1.** 7 **2.** 8 **3.** 12

**4.** 16 **5.** 17 **6.** 18

**7.** What kind of number is 22? 35? 41? 57? 71? 103? 261? 352?

**For items 8–11, complete this question:** What kind of number is the sum of:

**8.** an L number and a b number?

**9.** a b number and a rectangle number?

**10.** a rectangle number and an L number?

**11.** a b number and a b number?

### Extending Concepts

**For items 12–13, suppose the least of three consecutive integers is a b number.**

**12.** What kind of number is the middle number? the greatest number?

**13.** What kind of number do you get if you add the two least numbers? all three numbers? Is the sum of the three numbers divisible by 3? How do you know?

**In the 4-stacks model, cubes are stacked in columns with a height of 4 cubes. The rightmost column may contain 1, 2, 3, or 4 cubes as shown. Use the 4-stacks model to answer items 14–15.**

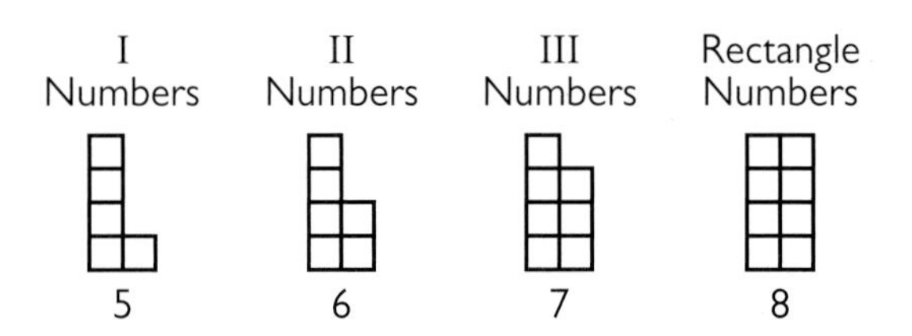


**14.** What kind of number do you get if you add a I number, a II number, a III number, and a rectangle number?

**15.** Is the sum of four consecutive integers always divisible by 4? Use your answer to item **14** to explain your answer.

### Making Connections

**16.** The *mean* of a set of numbers is found by adding the numbers and dividing by the number of numbers. Do you think that the mean of three consecutive integers is always an integer? Use what you have learned in this lesson to explain your thinking.

# A Stretching Problem

### Applying Skills

For items 1–6, remember that at the Bubble Gum Factory, 1-inch lengths of gum are stretched to lengths from 1 to 100 inches by putting them through a stretching machine. There are 100 stretching machines. Which of the following machines are unnecessary?

| | | |
|---|---|---|
| **1.** 5 | **2.** 14 | **3.** 19 |
| **4.** 31 | **5.** 37 | **6.** 51 |

For items 7–12, tell what necessary machines give these lengths:

| | | |
|---|---|---|
| **7.** 12 inches | **8.** 30 inches | **9.** 44 inches |
| **10.** 66 inches | **11.** 72 inches | **12.** 18 inches |

For items 13–15, find the prime factors of each number.

| | | |
|---|---|---|
| **13.** 18 | **14.** 48 | **15.** 75 |

**16.** How can you tell whether a number is divisible by 3? by 9? by 5?

### Extending Concepts

For items 17–19, use the same stretching machine described for items 1–12.

**17.** If you use only necessary machines, for which of the lengths 1 to 100 inches would exactly four runs through a stretching machine be needed? Give the lengths and tell which machines would be needed for each one.

**18.** Suppose a particular length can be obtained by using the necessary machines 2, 2, 3, 7. If you could also use unnecessary machines, what other combinations of machines could be used to obtain the same length? How did you solve this problem?

**19.** Describe a method to identify all the prime numbers between 1 and 200. Explain why your method works.

### Writing

**20.** Answer the letter to Dr. Math.

Dear Dr. Math,
We had to find all the prime numbers between 1 and 100. I think I noticed a pattern: All the odd numbers that were *not* prime were divisible by either 3, 5, or 7. So now I can tell whether *any* number is prime: If it's even, I eliminate it right away. If it's odd, I have to figure out whether it's divisible by 3, 5, or 7. If it's not divisible by any of them, then I know I've got a prime number. Has anyone else noticed this pattern? How should I explain why this is true?
In My Prime

# Pattern Appearances

## Applying Skills

Use the table you created in class or copy and complete the table below to show how many times each number from 1 to 25 would appear on a Multiplication Chart if the chart went on to infinity.

| 1 time | 2 times | 3 times | 4 times | 5 times | 6 times | 7 times | 8 times | 9 times | 10 times |
|---|---|---|---|---|---|---|---|---|---|
| 1 | 2 | 4 | | | | | | | |
| | 3 | | | | | | | | |
| | 5 | | | | | | | | |

For items 1–7, write down which column of the table each number belongs in.

**1.** 31 **2.** 26 **3.** 36 **4.** 41 **5.** 52 **6.** 58 **7.** 60

## Extending Concepts

Use your table to answer items 8–11.

**8.** In which column would 132 appear? Why?

**9.** Find three numbers greater than 100 which belong in the 5-times column. How did you find them?

**10.** What can you say about the numbers that appear in the 2-times column?

**11.** Is it possible for a number that is not a perfect square to appear in the 3-times column? Explain why or why not.

For items 12–13, suppose you listed all the numbers from 1 to 100 in your table.

**12.** How many columns would you need? Which number or numbers would appear in the rightmost column?

**13.** Which column would have the most numbers in it? Why?

## Making Connections

For items 14–15, use the following information:

The Ancient Egyptians had their own system for multiplying numbers. Examples of their system are as follows: To multiply a number by 2, double it. To multiply a number by 4, double it twice. To multiply a number by 8, double it three times. To multiply by 12, since 12 can be written as $8 + 4$, multiply the number by 8 and by 4 and add the results.

**14.** What are the prime factors of 8? Why does it make sense that to multiply a number by 8, you would double it three times? Use this method to multiply 11 by 8 and describe the steps.

**15.** How do you think the Ancient Egyptians would have multiplied a number by 14?

# The Final Arguments

## Applying Skills

Use the table you created in class or copy and complete the table below to show how many times each number from 1 to 25 would appear on the Multiplication Chart if the chart went on to infinity. Use your table to answer items 1–5.

| 1 time | 2 times | 3 times | 4 times | 5 times | 6 times | 7 times | 8 times | 9 times | 10 times |
|---|---|---|---|---|---|---|---|---|---|
| 1 | 2 | 4 | | | | | | | |
| | 3 | | | | | | | | |
| | 5 | | | | | | | | |

**1.** The perfect squares 1, 4, 9, 16, and 25 all appear in a column headed by an odd number. Do you think that *all* perfect squares belong in a column headed by an odd number? If so, explain why you think so. If not, give a counterexample.

**2.** What is special about the perfect squares that appear in the 3-times column? Explain why this makes sense.

**3.** Are there any numbers in a column headed by an odd number that are not perfect squares? Why or why not?

**4.** Do you think that all perfect cubes, other than 1, belong in the 4-times column? If not, give a counterexample. What is special about the perfect cubes that appear in the 4-times column? Why does this make sense?

**5.** Summarize the patterns regarding perfect squares and perfect cubes in an accurate mathematical argument.

## Extending Concepts

**For items 6–8, test whether each statement about cube roots is true.**

**6.** The cube root of any number is less than the original number if the original number is a positive fraction.

**7.** The cube root of any number is less than the original number if the original number is 0.

**8.** The cube root of any number is less than the original number if the original number is smaller than $-1$.

**For items 9 and 10, test whether each statement about divisibility is true.**

**9.** If you divide 10 by a positive fraction, the answer will be smaller than 10.

**10.** If you divide 10 by a negative number, the answer will be smaller than 10.

## Writing

**11.** Write a paragraph explaining the steps that are involved in making an accurate and complete mathematical argument. Explain why each step is important. Which special cases might you test? Why?

The McGraw·Hill Companies

This unit of MathScape: Seeing and Thinking Mathematically was developed by the Seeing and Thinking Mathematically project (STM), based at Education Development Center, Inc. (EDC), a non-profit educational research and development organization in Newton, MA. The STM project was supported, in part, by the National Science Foundation Grant No. 9054677. Opinions expressed are those of the authors and not necessarily those of the Foundation.

CREDITS: All photography by Chris Conroy and Donald B. Johnson.



Send all inquiries to:
Glencoe/McGraw-Hill
8787 Orion Place
Columbus, OH 43240-4027

ISBN: 0-07-866810-7

5 6 7 8 9 10 058 12 11 10 09 08